welcome little one!

Counting
MY FIRST NUMBERS

1

Counting

HOW MANY LEMONS DO YOU S

Counting

MY FIRST NUMBERS

Counting

HOW MANY GREEN LEMONS DO YOU SEE?

Counting

MY FIRST NUMBERS

Counting

HOW MANY APPLES

Counting

MY FIRST NUMBERS

Counting

HOW MANY FRUITS ARE IN THE BLENDER?

Counting

MY FIRST NUMBERS

5

How many?

How many?

3   5   4

7 Counting
MY FIRST NUMBERS

How many?

How many?

Counting
MY FIRST NUMBERS

How many?

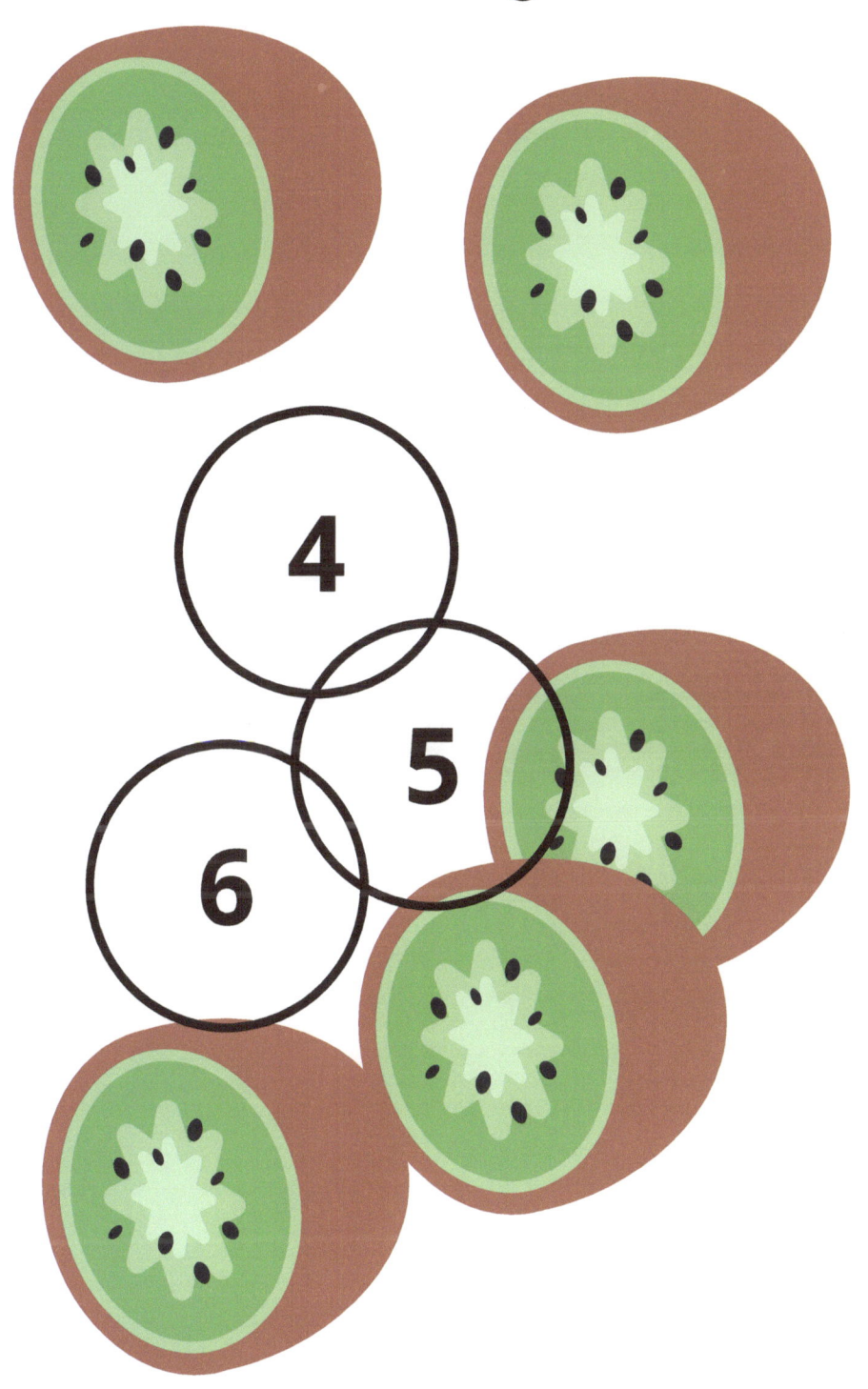

Counting
MY FIRST NUMBERS

How many?

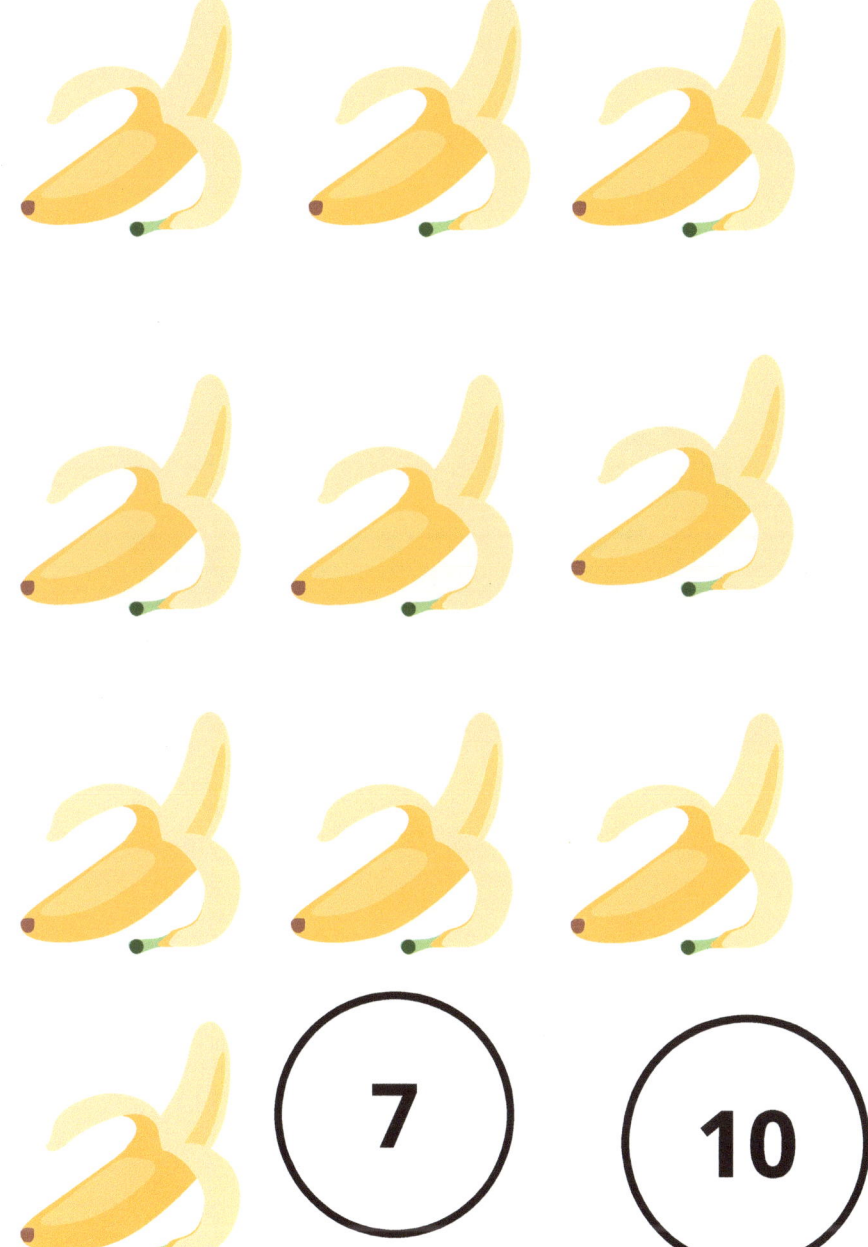

How many?

How many?

How many?

Colour

Colour

Colour

Colour

Colour

Colour

www.ingramcontent.com/pod-product-compliance
Lightning Source LLC
Chambersburg PA
CBHW041949240526
45473CB00036B/2787